Cheasapeake

Bay

Book

By. Autumn & Lucia

By: 4th grade

LIFE IN THE BAY

We dedicate this book to:

Ms. Bobbi

WATER CYCLE

WATER SHED

The Water Cycle

By: Eli Adams and Andre Smith

First the rain comes down onto the ground. It is called **precipitation**. Precipitation can be snow or rain. Precipitation makes surface runoff. Surface **runoff** is runoff that doesn't absorb in the dirt. The runoff goes in the river, then into the **streamflow**. Streamflow is the water that runs down a river or stream. When it rains the water in the stream gets higher. These streams flow into the bay and the ocean. The runoff that gets absorbed into the dirt, is called **infiltration**. Infiltration creates groundwater storage and **groundwater flow**. Groundwater flow is the water supply underneath the Earth's surface. Next, the sun's rays **evaporate** the water with the heat. The heat makes the water turn into steam or vapor and then it evaporates. It evaporates in the atmosphere and the air and then it rises. Then, when the evaporation comes together it forms a cloud and this is called condensation. **Condensation** means the changing of a gas to a liquid. The water cycle describes all the ways water can be and how water moves from one stage to another stage and how it keeps going. Water in the water cycle can exist as a solid, liquid or gas.

The WATER CYCLE

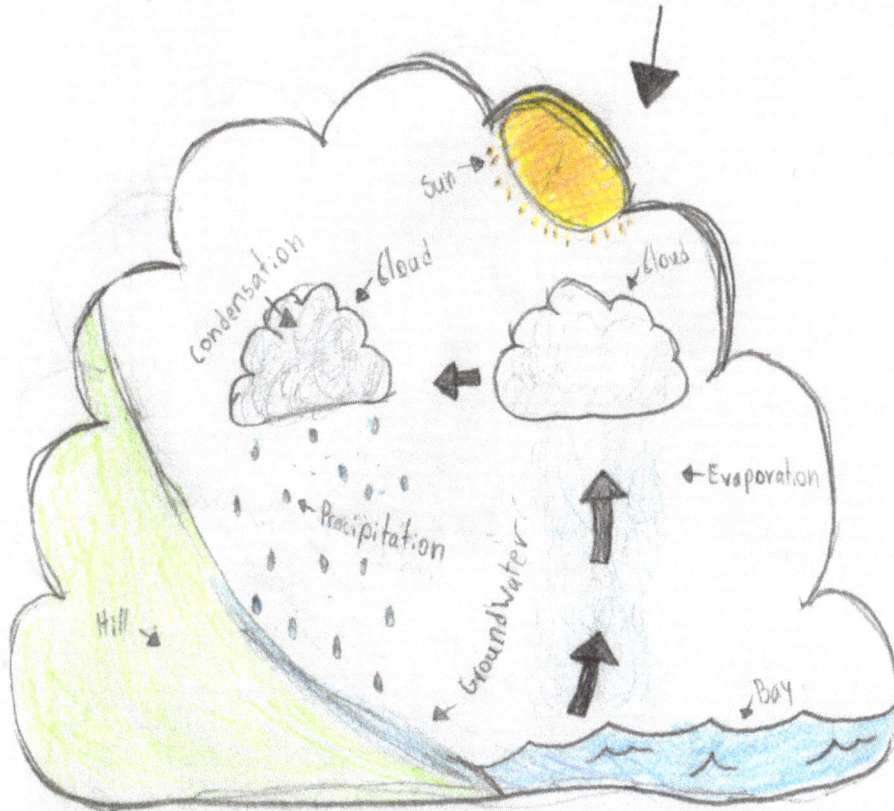

1. Evaporation: The trasformation of water to vaper.
2. Precipitation: Rain, Snow, Sleet, or hail falling to the ground.
3. Condenstion: The process of water vapor terning into liquid water.
4. Ground water: The water supply under the earths serface.

Chesapeake Bay Watershed
By Milan and Faylise

What is a watershed?

A watershed is many different rivers, streams, lakes, and reservoirs flowing into a huge body of water.

Fun Facts

- The bay can be stinky a lot of the time because the pollution ruins the smell of the bay. For example there is wastewater. Let's say that you are using the sink and when you put the water on whatever goes down the drain that is wastewater.
- Grass in the watershed absorbs runoff and rain.
- The Watershed helps the Chesapeake Bay stay cleaner though the grasses and muddy lands.

What states are in the watershed?

Delaware, New York, Maryland , Pennsylvania, Virginia, and West Virginia.

Brackish Water

by:Allen and Aidan

I am half salt

And I am half fresh.

My saltinest comes from the south

And my fresh water comes from the rivers.

I am salt water, I am traveling into the bay.

I am fresh water, I am already in the bay.

combined I am brackish water.

,

HABITATS, ORGANISMS!

Aidan And Allen's Wetland

I am sometimes dry

And I am sometimes wet.

My water Temperature is 18.1 celsius

and my turbidity is 5.9 NTU

That means it is clear.

Wetlands help slow the flow of nutrients, sediment and

chemical contaminants into rivers streams and the bay.

 Muskrats,Mallard,Canada geese,Wild celery,Broad -leaved cattail live in the

wetlands.

Muskrats are large rodents,

Canada geese are waterfowls and mallards too and

Wild celery are sav witch means submerged aquatic vegetation.

I am wetlands.

I am poem

By: Natalle

.

I am black and white
I eat fish, birds, and even crustaceans.
I live up to 30 years old.
I lay my eggs in late April
I can be found in South Antarctica
I swim in the water and I can fly
I am an Osprey

Fun Facts About an Osprey:

1. When an Osprey is about to lay their eggs they lay their eggs between April and May.
2. Ospreys will make their nest out of large sticks.
3. They live near rivers, lakes and ponds.
4. Osprey migrate from canada to south america every year fly 6,000 miles more
5. Osprey mostly eat fish but sometimes they eat

Mallard

BY GABE

I have a beautiful crown of green feathers on my head.
I can be found near lakes, marshes and other bodies of water.
I eat by sticking my head into the water and grabbing fish.
I only do this in water sixteen inches deep.

I have 8-12 ducklings a year and right after they are born they take a swim.
Our females takes feathers from herself to make a warm nest for her babies.
I can swim fly and find food one or two days after my birth.
I shed my feathers twice a year and that's called molting.
My outer feathers are waterproof because of a oil that comes from my tail.

I am the mallard duck.

Facts about the Oyster Catching Bird

By:Elijah Holmes

Oyster catching birds are most commonly found across the Atlantic coast line and have impressive orange/red beaks. They mostly feed on mussels. Also with it's tuxedo featured body their habitats are on the Sandy Grounds.

The male ones mostly do the nesting because they do the digging and make a small hole.

also 2 of the main breeds are Eastern and Atlantic.And the.they also they there eggs in scratches in the ground.their strong bills allow them to tear open there shell and tear through the mussels.

Fun fact:they dive down for the oyster because as adult oysters stick to one hard service and stays there.

so when they dive down to get it they grab it.

Habitats of Intertidal Flats

By.Angel

The definition of Intertidal is an area of land that is covered at high tide and uncovered at low tide.   A flat is a smooth surface and it not bumpy it can be but it mostly is a smooth surface. The Intertidal flats are habitats where certain sea animals can have shelter to call home.Today I will be talking about the animals that live in Intertidal flats and habitats that live in the bay. I also will talk about its predator and prey.

A big blue heron is a big bird that has a very strong beak and is very tall. A big blue heron is tall because of their legs. They have big wings. Their main source of food is fish.

Clam worms are five to six inches long, but near the bay they are not as long. The worm has two antennas and eight tentacles. It's not normal that they have atinas to sense things like predators and prey. Clam Worms eat clams and fish.

Soft shell clams. Their shell is nice and smooth. They are filter eaters which mean if any pollution is around they eat it like an oyster. The blue crabs eats soft shell clams.

 Diamondback Terrapin it eats clams. They call it that because they are strong. They also live in brackish shallow water in the bay. They are very strong clinger. They eat clams and crows. Birds also can eat them.

Diamondback Terrapin

Great Blue Heron

Clam Worms

Soft Shell Clam

D'aira's and Shynell's Trip to the Sandy Beach

By: D'aira and Shynell

Once a upon a time there were two girls on a trip to a sandy beach on vacation. They had their cameras and bathing suits and were looking forward to relaxing and swimming!

When they got there they put on sunscreen and ran down to the water. D'aira saw a Ruddy Turnstone, and that's when she got really happy! It had black and white spots on its head and throat.

"Wasn't that in our BayQuest?" D'Aira said.

"Yeah, you are right I always wanted to see one! They live in the sand." said Shynell.

"Let's explore the beach and see if we can find more!" said D'Aira. They were skipping and moving their arms as they went down the sand. They were barefoot and the sand felt soft on their feet. That's when D'aira stepped on a Ribbon Worm because it was in the sand and D'aira did not see it.

"Oh, what is that?!? Oh no I hope I didn't kill that animal." D'Aira said.

"Did you know they swallow their food whole?" Shynell asked.

"Ew that's gross?" D'Aira said.

D'aira and Shynell kept walking on the beach and saw a herring gull. "Oh D'Aira! There goes a herring gull! Did you know that a herring gull is one of the four types of gulls that lives in the Chesapeake Bay?"

"Oh, I didn't know that! I always knew they were white and gray but I didn't know their name." D'Aira said.

"Their bodies are about 20 inches long!" Shynell said as she pointed at the gull.

"Wow we learned a lot about the bay this year. I hope we learn more." D'Aira said while she walked along the waters edge in her bare feet.

BLUE HERON BLUE HERON

By: Darntrell Boyd

I HAVE A SMALL BODY WITH A LONG NECK AND LONG FEET

I CAN STAND ON ONE FOOT

I CAN BUILD NESTS FOR ME TO LIVE IN

I Can Fly

I CAN CATCH FISH WITH MY LONG BEAK

I CAN EAT DIFFERENT TYPES OF ANIMALS

I AM A CARNIVORE

I CAN EAT CRABS AND FISH

I AM A BLUE HERON

Great Blue Heron

By: Milan

I'm a bird as *beautiful* as the night sky.
Hunting for *food* to *survive*, even though I can't *fly*.
I stand in the *water*, *fish* hiding because they are *shy*.

Ducking my head in the *water*, looking for fish to *eat*.
I found one, and I start devouring its *meat*.
I look like a *swan*, just *blue* at some parts.
Being scared of *bobcats* and others, scared to be
eaten.
Even though they prefer me *sweeten*.
I am a great blue heron.

This Is The Aquatic Reefs

By Liam and Jeremy

    While I'm swimming,
I see fish gathering around spiky red bearded sponge's,
I see algae blooms and beautiful underwater plants,
This is the Aquatic Reefs.

    While I'm swimming,
I see a Eastern Oysters camouflaging its gray shell with the rocks,
I see the slug-like oyster drills coming out of its shell searching for shelled objects for food,
This is the Aquatic Reefs.

    While I'm swimming,
I see the Sea Squirts with its 2 or 3 tubes flowing by the movement of the water,
I see the dirty sediment filled water washing from lawns to streets to the bay,
This is the Aquatic Reefs.

    While I'm swimming,
I see the dirty brown rocks with grey oysters laying on them,
I see the filth sand with little rocks and stones covered inside of it,
This is the Aquatic Reefs

    While I'm swimming
I see the ragged, brown, oyster eating Toadfish rising from its habitat we call Aquatic reefs.

The Oyster Drill Poem

By:Nick Z

I am gray or purple and sometimes tan.

My favorite things to eat are oysters and clams.

I have a hard shell to protect me from predators.

I am scared of blue crabs, birds, and fish.

I can smell an oyster's body odor to tell if i can eat them.

You can find me in the bay grass beds.

I am the oyster drill.

Oyster Drill

Red beard sponge

By: Jordyn Robinson

I am a red beard sponge.
Sponges like me are sometimes are orange and others are red.

I am a filter feeder.
Sea slugs,starfish and turtle eat me
And I give shelter to shrimp, mud crabs,worms and other small fish live in me.
Fun fact some people think I am a plant
But I am a animal.
 Another fun fact I can't survive out of water
And I have many cells to help me survive.
And my final fun fact is I can recreate myself.
For example, if you squish me
 I can find different cells and eventually I am recreated.
I attach to any hard surface.
 For example, some places you can find me in are
Oyster beds and rocks throughout the bay.

This is The Bay Grass Beds

By Liam and Jeremy

While I'm canoeing,
I see the turfy green grass waving around in the water,
I see the grains of gray rough rocks ,
This is the Bay Grass Beds.

While I'm canoeing,
I see the weak, jagged Widgeon Grass fluttering,
I see the ferm, solid shell of a Grass Cerith coming out of its shell,
This is the Bay Grass Beds.

While I'm canoeing,
I see the male Seahorse carrying its baby around the water,
I see the spiky feeling of the Blue Crab,
This is the Bay Grass Beds.

While I'm canoeing,
I see the smooth, brown Cownose Rays,
I see the green plants layered all over the dirt,
This is the Bay Grass Beds.

While I'm canoeing
I see the wavy Eelgrass covering the habitat called Bay Grass Beds.

Blue Crabs

By: Jaiden

ABOUT THE BLUE CRABS

*1.*Blue Crabs are a crustacean with bright blue claws and a olive green shell.

*2.***Blue Crabs** live in Bay Grass Beds.

*3.*A Blue Crabs shell varies in color from bluish to olive green, and can reach up to 9 inches across.

*4.*Blue Crabs eat clams, oysters, mussels, smaller crustaceans, freshly dead fish, and plant and animal detritus is a waste of any kind, and they eat smaller and soft-shelled **Blue Crabs**.

Predators

The animals that eat Blue Crabs are Croakers, Red-Drum, fish-eating birds like The Great Blue Herons, sea turtles and jellyfish.

Reproduction and Life Cycle

Blue Crabs can mate from May to October in the brackish water. The way the **Blue Crabs** mate is by, The male crab carries the Female Crabs and takes her to an area and then mates. When they're done, the male crab takes the female crab again and mates in a different area. When the male crab is done, The female crab's shell will start to harden while the male crab leave's to mate with another female crab. The female crab will lay **750,000** and two million eggs.

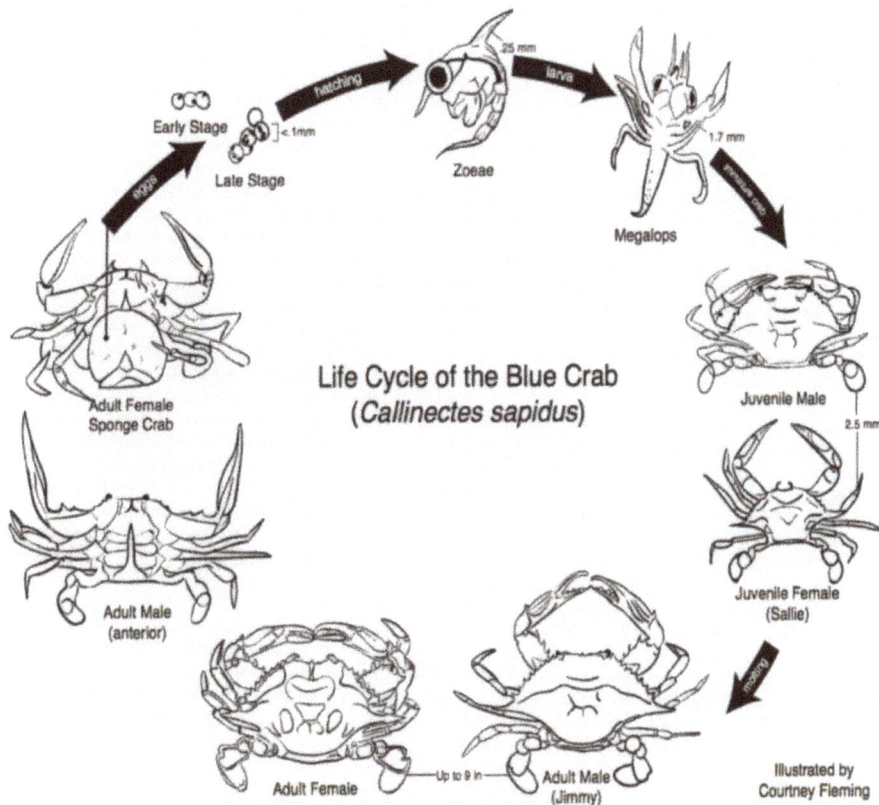

Life Cycle of the Blue Crab
(*Callinectes sapidus*)

Early Stage

Late Stage

< 1mm

hatching

Zoeae

25 mm

larva

1.7 mm

Megalops

Juvenile Male

2.5 mm

Juvenile Female
(Sallie)

Adult Female
Sponge Crab

Adult Male
(anterior)

Adult Female

Up to 9 in

Adult Male
(Jimmy)

Illustrated by
Courtney Fleming

The stages for the **Blue Crabs** are called:
1. Baby/Eggs
2. Hatching:Zoeae
3. Larva:Megalops
4. Immature crab:Juvenile Male/Female
5. Adult Male/Female
6. Adult Male (anterior)
7. Adult Female sponge crab

Blue Crab

I'm know to be the symbol of Maryland, known to be a NUTRITION.

Even though i'm ENDANGERED. I do have a foundation.

I have relation to regular crabs, not to be bragging.

Fisherman are mean, trying to overfish me, I'm pretty sure they are nagging.

I've grown from a plankton, sure to be eaten by oysters makes me go gagging.

I am the blue crab

COWNOSE RAY

By: Ben Q

I am brown like a bear.
I have a tail like a whip.
I am an opportunist feeder,
I get what I get and I eat it all.
I am an open ocean species, I'm everywhere!
I have flat teeth to crush my shelled prey.
I am known for my long distance migrations.
Also, please don't kill me for " contests".

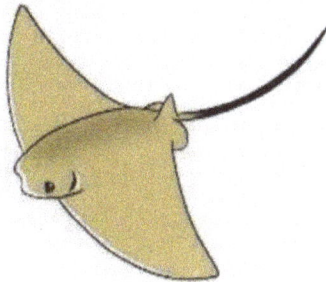

I am the Cownose Ray.

I Am A Seahorse

By:Autumn

I am green, sometimes I can change colors when I am

in a mood or in love.

I get shelter from eelgrass.

I eat eat many things such as plankton,clam worms.

I don't have a lot of predators because I am to boney.

I live in the bay grass beds.

I breath through my gills and have a swim bladder.

I live up to 10 years.

There are 40 species of me.

You can keep me as a pet.

My name has been given to 54 species of small marine fish.

I can deliver more than 1,000 babies at a time.

I prefer to swim with my tail on my friends tail while I

am swimming.

I am a seahorse.

Eelgrass

Some people pull me from the ground.
I am made up by long ribbon leaves.
My leaves can grow to 4 feet.
The black duck feeds on me.
I grow near the shore.
I'm mostly known as Zostera Marine.
I can be found in bay grass beds.
I make my food from the sun.

I am eelgrass

I am seahorse poem

I can change colors when I am in love or camouflaging.

I live in the aquatic reefs and bay grass beds.

I don't have many predators because I am too boney.

Some of us only live up to 10 years.

I eat plankton and small fish such as shrimp.

We prefer to swim in pairs with our tails linked together.

I am named for the shape of my head.

I have a long tail like a snake.

I have a swim bladder.

There are 40 known species of us.

We don't have teeth we actually suck our food in whole.

Our fins flutter up to 35 times per second.

We can consume 3,000 or more brine shrimp per day.

I am a seahorse

By: Lucia Guevara

This Is The Deep Open Waters

By Liam and Jeremy

While I'm scuba diving,
I see silverish Striped Bass soaring through the dark blue water,
I see brown Cownose Rays flying through the coral,
This is the Deep Open Waters.

While I'm scuba diving,
I see groups of fish swimming through the daylight of green water,
I see grey Menhaden skimming the seaweed,
This is the Deep Open Waters.

While I'm scuba diving,
I see the deep, deep bay with detailed shells at the bottom,
I see sharp nosed Sturgeons increasing its speed,
This is the Deep Open Waters.

While I'm scuba diving,
I see sea Cucumbers just lying there,
I see sea nettles stripes on their back on their back,
This is the Deep Open Waters.

While I'm scuba diving,
I see sharks squirming around,
But I'm not scared,
Because I'm in the Deep Open Waters.

Striped Bass

By: Aminah

I have four nostrils.

I can grow to about 2 to 3 feet in length.

I am a popular fish in the Chesapeake Bay.

The Bay is a big nursery area for me.

I'm Maryland's state fish.

I have to eight to seven stripes on me.

I usually live up to 30 years.

My growth can depend on where I am.

I can grow to 55 to 77 pound in weight.

I am striped bass

Striped Bass Poem

By Caleb

I am silver with dark stripes on my back and that's how I got my name.
I am the state fish of Maryland.
I can be up to 125 pounds but I am usually 67 pounds and 8 ounces.

I've been 31 years old.
I eat menhaden, sea herring, flounder, silver hake, squid, and more.
I am native to the chesapeake bay.
I am a Striped Bass.

Sea Cucumber

By Aminah

I have about 1,200 known species of me.
I eat plankton and algae.
I have been used for medicine in Asia.
I am a hiding place for pearlfish.
I can be as small as 0.12 inches.
I can be as big as 3 feet long.
I am sea cucumber

I am poem

By: Andre and Eli

I am a long fish that lives in **deep open water.**
They have a long rubber noses.
They eat **clams**, **worm**, **snails**, **crabs**.
They have no teeth.
They are eaten by **sea lampreys** and **sharks**.
The top of their tail is longer than the bottom.
Their color is gray or brown.
They are bottom feeders.
Bottom feeders are animals that eat on the bottom of the bay.
Also they search for their food.
In the old time they were important part of the fishing trade of the Chesapeake Bay.
They live up to 100 years.
Guess what it is called.
It's called a **sturgeon.**

THE STURGEON

By Julian

I am big and silver darting this way and that

I can be 2,500 pounds but I'm not fat.

I'm native to the bay but I am endangered.
I'm a prehistoric fish that has been in the bay
For millions of years.I am almost invisible when
I'm an adult because of my size, I am the Atlantic
Sturgeon.

The Story of the Oyster's Life Cycle
By: Autumn and Natalle

This is how oyster are made:

When male oysters release sperm the females oysters can release 5 to 8 million eggs at one time. They then create fertilized eggs. Once the egg is fertilized, the egg grows a little bit. When the egg develops it becomes a veliger. At this age the veliger is growing but still young so they are still moving. With further development, the pediveliger grows a foot to get ready to stick in one place. A pediveliger is still young so they are still moving. A spat is when they are growing, but they have found a place to live so they are stuck in that place. Finally the oyster is fully grown and they are living in one place cleaning the bay.

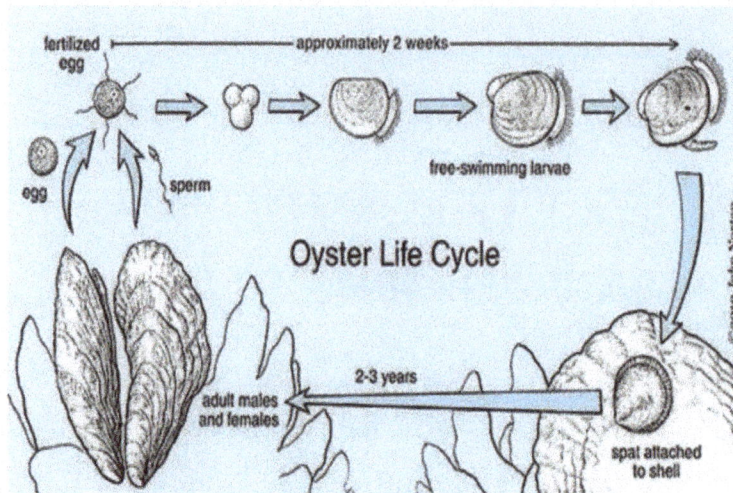

approximately 2 weeks

fertilized egg

egg

sperm

free-swimming larvae

Oyster Life Cycle

©2010 John Norton

2-3 years

adult males and females

spat attached to shell

Fun facts

→An oyster has to be 3 inches in order for a water man to collect them.

→Another fun fact is oysters grow up to one inch year.

→Also another fun fact is that the biggest oyster is 14 inches.

→Also we need to stop overfishing them, that way the oyster will stop dying from people overfishing them.

→We need oysters to help us clean the bay but they keep dying from water men and msx which is a disease that oyster keep catching. Don't overfish oyster!

Oyster's Food Chain

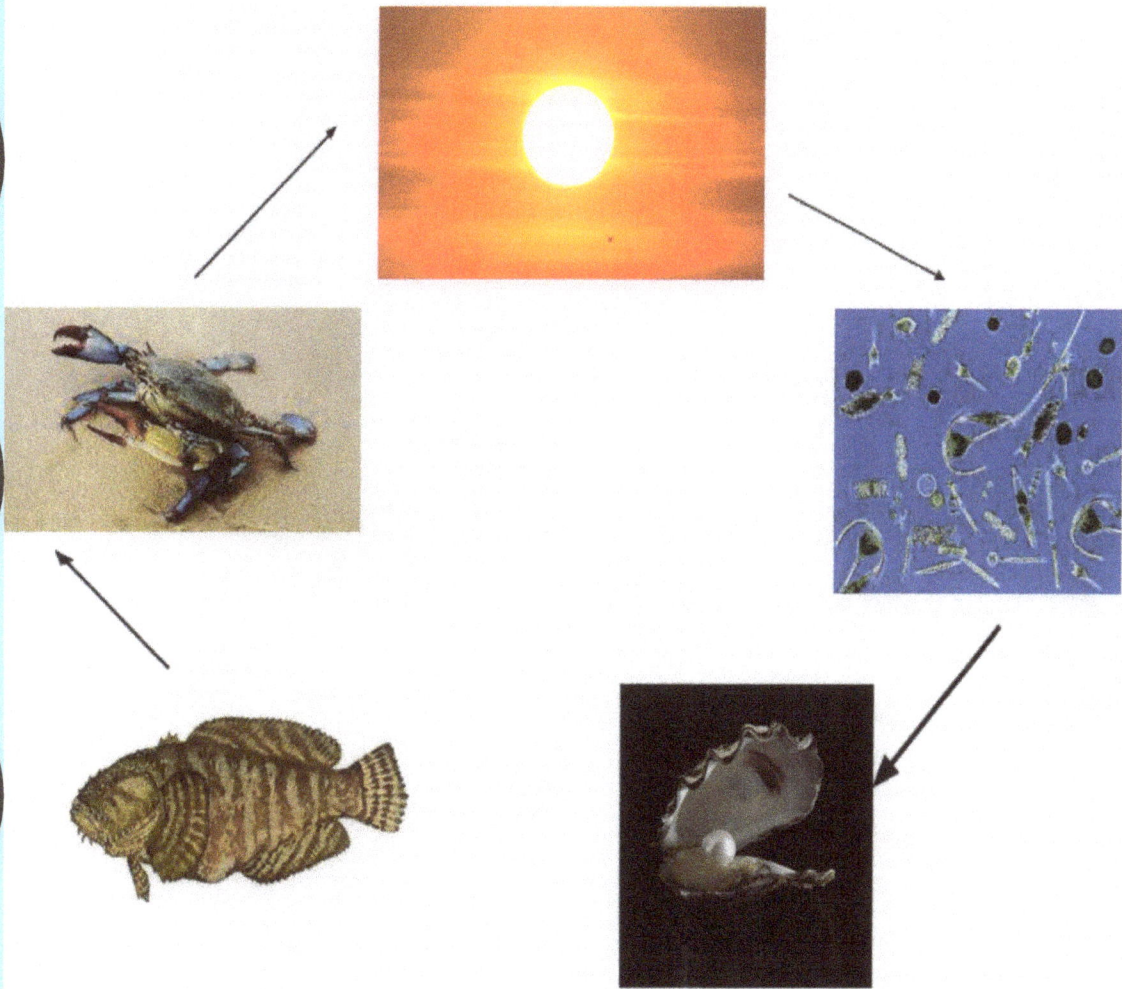

The "Oyster's Food Chain" is about different animals in the bay. The Chesapeake Bay that need each other to survive. phytoplankton~ For example, the sun gives the sunlight to make their own food. Then, the oysters eat the phytoplankton. next the Oyster Toadfish eats the oysters. Lastly, when the oyster toadfish dies, the crab eats the toadfish dead body this is the oysters food chain.

1. Sun: The sun start's off the photosynthesis by giving the plants and animals in the bay sunlight.

2. phytoplankton: The sun gives them sunlight so they can make their food through photosynthesis.

3. oyster: Oysters are an animal that clean the bay by filtering the water

4. oyster toadfish: oyster toadfish can grow up to 12 inches and sometime oyster toadfish can mosley small crabs

5. blue crab: The blue crab has a 9 inch shell and it is blue and has 9 on each side also males don't have colorizition on the tips of their claws and females do.

photosynthesis: Means that animal or plant makes their own food.

colorizition: Means that they have color different colors all over them.

I am Phytoplankton

By:Shynell

Me, the small thing that doesn't get noticed,

But I live in the Chesapeake Bay.

I get eaten by oysters.

I convert sunlight into food.

I am one species with many different shapes.

I can reproduce 100 million offspring in a month.

You need a microscope to see me.

I create 80-160 billion tons of carbohydrates per year.

I am phytoplankton.

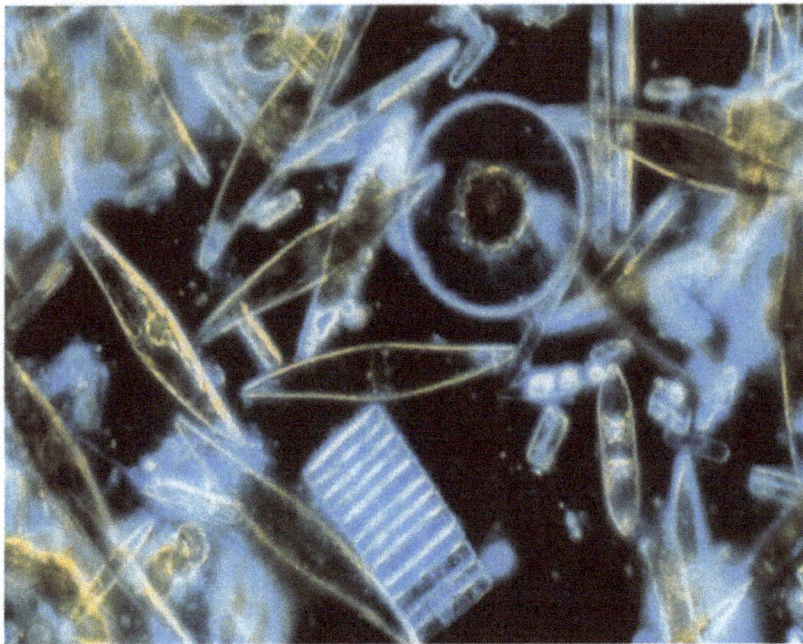

Oyster Poem
By: Jordyn Robinson

I am the one who filters the bay.

I love to eat **phytoplankton** and **toadfish** love to eat me.

You might be thinking how do I filter the bay?

Well here's your answer

I suck up a lot of water and I divide it in two different sections

One of the sections is food

And the other section is the dirt in the bay I sucked up.

However I spit out all the water I sucked up.

A few moments after waiting for all the dirt to clump up, and I form a mucus ball.

After I spit out the mucus ball and because it's so heavy

it falls right down in the bottom of the bay.

But now is to hard for me to filter the bay because

Water men overfish me.

For example it use to take me **a few weeks**

but now it takes about **3 to 4 years**.

I have a few different names in my cycle.

When I am a baby I am called veliger larvae

When i am a child I am called pediveliger larvae

But I can't tell you what my name is when I am full grown yet.

And when I am too young to attach to something or someone I am called **spat**.

You can find me in stores and a lot of people love to eat me.

I have many diseases and one is called **MSX.**

MSX is a strong disease that kills pediveliger larvaes.

And to end this I´ll tell you who I am I am a **Oyster**.

Oyster Poem

by.Angel

I'm a certain thing

that can clean like a vacuum

sucking up all the dirt I

stay in one place like on a hard surface

I have something protecting me like a shell

I eat phytoplankton I die of a certain kind of pollution if I am a larvae

I am something that could die easily by being overfished

I am something that needs help by not being extinct

I am something that makes pearl spit balls like a clam but my spit balls are made of dirt and spit

I can be eaten by something called a toad fish I only travel as a baby but when I grow up I stay in one place

I am something that stays in the water I can only clean my area not the whole bay I can be eaten by humans and sea creatures My heart and guts are inside my shell

I am something that can die in one second I need help surviving even from my predators that's how much I need help I am a certain animal that can save the food chain

I am something that can make fisherman loosie their job

I am something that not be pulled from my territory

Oyster Toadfish

By Faylise

I am a fish.
I have a snapping jaw and a sharp spine.
I am an Oyster Toadfish

I am an ugly fish.
You can find me around a lot of Oyster Reefs.
I can grow up to 12 inches in length.
I am an Oyster Toadfish.

Small crabs eat me.
I am an Oyster Toadfish.

The litter and polluted waters don't bother me
I can survive out of water for a period of time.
I am an Oyster Toadfish.

Facts About the Food Web

By Autumn and Jordyn

Do you know what a food web is? Well I can tell you what it is. A food web is a food chain, but it goes a different way. It's like a web, connecting all the different food chains.

Do you know how a food web is formed? Well, it is formed by the circle of life and the food chains. First an animal eats a plant, then another animal eats that animal that was eating the plant and it creates a food chain. When a lot of animals are in a food chain they create a food web. For example, phytoplankton are filter feeders and seahorses eat plankton. Also oysters eat phytoplankton. Then the oyster toadfish eats the oysters and crabs eat the seahorse and the oyster toad fist.

The food web is so im `portant because when these animals eat the plant they produce food energy and organic compounds.

How does a food web work?A food web cooroperates by a food chain because they are kind of the same but not the same. One animal eats an animal and the animal that gets eaten eats something that gets eaten by another animal. This is how a food web works.

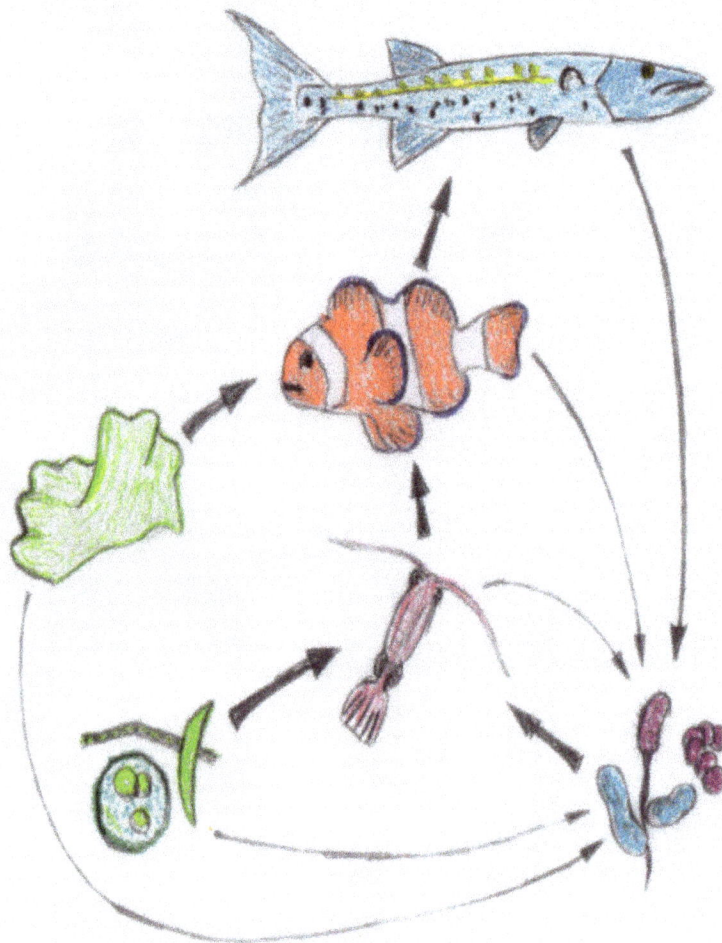

Runoff

By: Gabe McMahon and Caleb Haber

Runoff is rain that comes down and does not get absorbed. It keeps going and doesn't get absorbed by impervious surfaces. Impervious surfaces are man-made areas that cannot absorb water from rain or snow. The runoff picks up trash, wastewater (water that has been used), and chemicals and then takes the pollution to the bay. Runoff also picks up sediment, in this case sediment is dirt that clouds the water to hurt the bay. To stop this from happening you can: limit the amount of fertilizer you use, pick up trash if you see it on the ground, and plant native plants because they help keep runoff from getting into waterway.

Fertilizer

I am on the grass for growing it.

I am full of nitrogen, phosphorus and potassium.

I am being knocked of the grass.

I am moving with the rain.

I am continuing to be pulled by the rain.

I have been left on the street. All I can see is a big mass of water.

The rain is now leading me into the water.

I fill the bay with my nutrients.

I make algae blooms with my phosphorus.

My potassium deteriorates the water quality of the bay.

Like my potassium, my nitrogen deteriorates the water quality.

I am the fertilizer on your grass.

Waste Water

I am smelly. I come from your toilet.

I live in the sewer. I go through the pipes but then they BURST!!!

I am going around town, up and down. Then I go into the bay.

I put nitrogen in the bay and deteriorate the water quality.

The phosphorus creates dead zones, no fish and no plants.

I am your toilet wastewater.

Facts About Mr. Trash Wheel!

By: **Jaiden & Elijah**

1. Mr. Trash Wheel is a machine that cleans up the bay by picking up trash from runoff to save the bay from pollution!

2. Mr. Trash Wheel is a machine that only runs on solar power and hydropower. Solar power is the sun and hydropower is hydroelectric power, that uses the movement of water.

3. Mr. Trash Wheel can be found where The Jones Falls Expressway meets the Chesapeake Bay

4. When the Trash Wheel is full, a boat will take it to a dumpster call "Volia`".

5. In four years, Mr. Trash Wheel collected 1,510,620 pieces of trash.

6. The workers at the Chesapeake Bay collected all the cigarette butts and lined them up, and it added up to 70 miles!

7. Also, there is another trash wheel called captain Trash Wheel" that is in progress right now.

Mr. Trash wheel   Professor Trash Wheel

More Plants

By: Amara Coughlin

We should have more plants.

Supporting detail # 1 We need more plants "to be safe."

Supporting detail # 2 We need plants "stop the water."

Supporting detail # 3 We need more plants to "make more flowers."

Throw Out Garbage

We should throw out our garbage.

Supporting detail # 1 We need to throw out our garbage because "it keeps the bugs away."

Supporting detail # 2 We need to throw out our garbage " so that we can keep the streets clean."

Supporting detail # 3 We need to throw out our garbage so that " we can go outside and play."

www.ingramcontent.com/pod-product-compliance
Lightning Source LLC
Chambersburg PA
CBHW050241220326
41598CB00047B/7469